G000036716

FINANCIAL INSTRUMENTS TO STRENGTHEN WOMEN'S ECONOMIC RESILIENCE TO CLIMATE CHANGE AND DISASTER RISKS

AUGUST 2022

ASIAN DEVELOPMENT BANK

Creative Commons Attribution 3.0 IGO license (CC BY 3.0 IGO)

© 2022 Asian Development Bank
6 ADB Avenue, Mandaluyong City, 1550 Metro Manila, Philippines
Tel +63 2 8632 4444; Fax +63 2 8636 2444
www.adb.org

Some rights reserved. Published in 2022.

ISBN 978-92-9269-647-4 (print); 978-92-9269-648-1 (electronic); 978-92-9269-649-8 (ebook)
Publication Stock No. TCS220335-2
DOI: http://dx.doi.org/10.22617/TCS220335-2

The views expressed in this publication are those of the authors and do not necessarily reflect the views and policies
of the Asian Development Bank (ADB) or its Board of Governors or the governments they represent.

ADB does not guarantee the accuracy of the data included in this publication and accepts no responsibility for any
consequence of their use. The mention of specific companies or products of manufacturers does not imply that they
are endorsed or recommended by ADB in preference to others of a similar nature that are not mentioned.

By making any designation of or reference to a particular territory or geographic area, or by using the term "country"
in this document, ADB does not intend to make any judgments as to the legal or other status of any territory or area.

This work is available under the Creative Commons Attribution 3.0 IGO license (CC BY 3.0 IGO)
https://creativecommons.org/licenses/by/3.0/igo/. By using the content of this publication, you agree to be bound
by the terms of this license. For attribution, translations, adaptations, and permissions, please read the provisions
and terms of use at https://www.adb.org/terms-use#openaccess.

This CC license does not apply to non-ADB copyright materials in this publication. If the material is attributed
to another source, please contact the copyright owner or publisher of that source for permission to reproduce it.
ADB cannot be held liable for any claims that arise as a result of your use of the material.

Please contact pubsmarketing@adb.org if you have questions or comments with respect to content, or if you wish
to obtain copyright permission for your intended use that does not fall within these terms, or for permission to use
the ADB logo.

Corrigenda to ADB publications may be found at http://www.adb.org/publications/corrigenda.

Note:
In this publication, "$" refers to United States dollars.

On the cover: (From left to right) **Photo of girls at bank.** Banco Nacional de Comércio de Timor-Leste (BNCTL) bank
client at the main bank branch in Dili, the capital of Timor Leste, the Expansion of Financial Services Project will support
the commercial transformation of BNCTL (photo by Luis Enrique Ascui). **South Asian girl.** Women farmer on farm field
(photo by Shutterstock). **Girl in red.** Loans from Access bank has enabled small businesses a chance to develop and
grow. As businesses ramp up operations, job opportunities are offered to women and other members of the community
who are socially disadvantaged (photo by ADB).

CONTENTS

TABLES, FIGURES, AND BOX

ACKNOWLEDGMENTS

This publication was prepared under an Asian Development Bank (ADB) technical assistance project (TA 9660): Promoting Transformative Gender Equality Agenda in Asia and the Pacific and funded through the Urban Climate Change Resilience Trust Fund (UCCRTF) financed by the Rockefeller Foundation and the Governments of Switzerland and the United Kingdom. It is part of a set of deliverables geared toward operationalizing the transformative gender agenda and increasing women's participation and leadership in decision making within urban governance systems in four project cities: Kushtia and Faridpur in Bangladesh, and Coron and El Nido in the Philippines. This publication particularly focuses on women's economic empowerment and resilience to external shocks.

Malika Shagazatova, social development specialist (gender and development) with ADB's Sustainable Development and Climate Change Department (SDCC), provided overall guidance in the development of the publication with support from Alih Faisal Pimentel Abdul, TA coordinator. Ma. Celia A. Guzon, senior operations assistant, and Noel Chavez, senior operations assistant (gender), both with SDCC, provided administrative support. The publication was edited by Jill Gale De Villa, and graphics and layout were done by Rocilyn Locsin Laccay and Prince Nicdao.

Ma. Piedad Geron conducted the original research and prepared this publication, with support and contributions from Ephrat Yovel; Resources, Environment, and Economics Centers for Studies, Incorporated (Philippines); and Services and Solutions International Limited (Bangladesh).

The publication benefited from peer reviews by Kelly Hattel, senior financial sector specialist, in the Southeast Asia Department; and Lisette Cipriano, senior digital technology specialist (financial technology services) in SDCC.

Special thanks to Samantha Hung, chief of gender equality thematic group in SDCC, for her overall support and guidance in the implementation of this technical assistance project.

ABBREVIATIONS

ADB	Asian Development Bank
CARD	Center for Agriculture and Rural Development
CCRIF	Caribbean Catastrophe Risk Insurance Facility
COVID-19	coronavirus disease
CPMI	CARD–Pioneer Microinsurance Incorporated
ELF	Emergency Liquidity Fund
FSP	financial service provider
IPCC	Intergovernmental Panel on Climate Change
MFI	microfinance institution
MSMEs	micro, small, and medium-sized enterprises
NFIS	national financial inclusion strategy
NGO	nongovernment organization
PKSF	Palli Karma Sahayak Foundation

EXECUTIVE SUMMARY

Disasters brought about by natural hazard events and extreme weather conditions affect men and women differently. Differences in socioeconomic status, ability to access resources, and prevailing social and cultural norms between men and women impact the way they respond to and are affected by disasters. Given the low rate of women's participation in the labor force and that women are mostly self-employed in the informal economy, they are more likely than men to live in the poorest households. With frequent exposure and high vulnerability to shocks, and with less ability to cope and recover from them, women may have limited ability to withstand the impact of disasters that cause financial shocks and economic distress. **Women's economic resilience to disasters is a result of how well women are prepared for and able to cope with such events.**

Financial inclusion is a state in which everyone, especially vulnerable people, has effective access to a wide range of financial products and services (savings, credit, insurance, remittances, and payments). Financial inclusion enhances women's economic resilience to disasters by providing them access to various types of financial instruments. Use of relevant and appropriate financial instruments enables women to prepare for and cope with the effects of disasters. An important element of financial inclusion is financial literacy—which is achieved by providing individuals with the knowledge, information, and tools that enable them to make informed financial decisions that in turn can prepare them for unexpected financial shocks brought about by disasters.

Financial inclusion enhances an individual's or household's level of preparedness for disasters. By using credit, savings, and/or insurance, individuals, households, and businesses can anticipate, prepare, and eventually absorb the negative impacts of a disaster by creating assets, smoothing consumption, and accessing emergency assistance. **Financial inclusion boosts an individual's or household's ability to cope during and after a disaster.**

When an individual has an account in a financial institution, cash transfers from government and remittances from family and friends can immediately be facilitated in the event of a disaster. Also, access to credit facilities can provide the resources needed for recovery and reconstruction after a disaster, making it possible to rebuild faster and better. Savings and proceeds from insurance claims may also be used to finance restoration, rebuilding, and recovery-related activities. Immediate access to funds (either through emergency credit or savings) helps smooth income and consumption right after a disaster. **Digital financial services play an important role in financial inclusion and enhancing economic resilience to disasters.** Digital financial services provide a cost-efficient way to reach and provide financial assistance to the most vulnerable people.

Recognizing the importance of financial inclusion for poverty alleviation and inclusive growth, various countries developed and implemented their own national financial inclusion strategies. Countries that have adopted a strategic policy approach to financial inclusion have contributed to an improved state of financial inclusion worldwide. The proportion of the global adult population reported as unbanked declined from 49% in 2011 to only 31% in 2017. However, gender disparities in account ownership persist, with male account ownership at 72% versus female account ownership at 65%. The 2017 Global Findex reported that the 7-percentage point gender gap has not changed since 2011 among developing economies. Despite the importance of financial inclusion for women, they still comprise 56% of all unbanked adults.

Women face multiple barriers to accessing and using financial services. Most of the barriers are a result of sociocultural practices and societies' perceptions of women, but some barriers are institutional or merely infrastructure related. The latter include women lacking an identification (ID) card to prove their identity, having insufficient or no traditionally required collateral, facing mobility constraints, and having limited financial literacy. Other barriers to women's account ownership include their concerns about the cost of an account and perceptions of the utility of accounts, a lack of sex-disaggregated data on the use of financial products, a lack of financial products that cater to women's needs, a lack of credit history, and the high costs of dealing with financial service providers (FSPs).

Women are often more vulnerable than men to disasters because basic services and assistance are "gender neutral" and do not consider unequal gender roles and discrimination women face in many societies. In times of shock or disaster, men's and women's capacity to access emergency funds may differ. Men are more likely than women to be able to access emergency funds. During emergencies, men use their earnings from work as a source of emergency funds while women tend to rely more on savings and money received from family and friends. Gender-responsive financial mechanisms that enable access to financial services reduce women's vulnerability, strengthen their capacity to cope with natural hazard events, and help ensure that they are not left behind when economies rebuild.

Relevant financial instruments may be used to increase women's level of preparedness and to enhance their abilities to cope with shocks. Financial instruments that increase a woman's level of preparedness include those that enable women to reduce risks, be prepared for disasters, and transfer such risks to insurance companies. In the event of a disaster, financial instruments are also used for relief and recovery measures. For women to access the financial instruments, they should have features that address specific gender barriers.

Risk reduction. To reduce disaster risks that women face, loan products for acquiring business and household infrastructure that reduce such risks could include the following key features:

(i)   designed for business expansion and/or livelihoods that will enable women to earn additional income and generate surpluses for precautionary savings in times of disaster;

(ii) finance climate adaptable infrastructure and/or equipment as collateral;

(iii) accept personal property as collateral for loan products that reduce disaster risks;

(iv) base loan repayment terms on women's cash flows and anticipate the occurrence of weather hazards, to minimize default losses;

(v)   provide simple documentation requirements; and

(vi)   disburse funds rapidly, with an option to release them to women's e-wallets.

Risk preparedness. Women tend to be more risk averse than men. Hence, to prepare for future disasters, women prefer to use their savings instead of relying on borrowed capital during emergencies. Women are also more inclined than men to use their savings for business expansion or investments. Hence, women prefer savings products and consider that such products are important for strengthening their economic resilience. Women consider precautionary savings, accumulated over time by regularly setting aside small amounts over time, as low-cost sources of liquidity. Research shows that even low-income women can save about 10%–15% of their net monthly income. Key features of savings products appropriate for women include (i) being goal-oriented—i.e., tied to a specific commitment goal and withdrawable only when the goal is met; (ii) accounts that may be opened with small amounts and simple documentation; (iii) accessible digital savings, with cash-in–cash-out modes that are available in channels that are familiar to and near women clients (e.g., convenience stores, rural agents); (iv) include training programs for women savers; and (v) do not impose fees and charges for maintaining account balances.

Risk transfer. Unlike idiosyncratic risks (i.e., risk events limited to an individual and/or household), which may be addressed by precautionary savings, "covariant" or catastrophic risks brought about by natural hazard events such as earthquakes, floods, and typhoons are best addressed through risk transfer or insurance. Insurance may be provided directly or indirectly. Direct insurance is offered directly to individuals through distribution channels such as microfinance institutions and cooperatives. Important features to consider when designing direct insurance products for women are premium affordability, use of distribution channels that are familiar to women, and bundling insurance with other financial products such savings and/or credit. With indirect meso insurance, groups or institutions that directly work with and for women (e.g., microfinance institutions, nongovernment organizations assisting women, and cooperatives with women membership) are the policyholders. These organizations then determine how the insurance claims will be distributed to their members. This type of insurance is a good start to orient women on the important benefits of insurance. Insurance literacy is an important factor to consider when designing insurance products for women.

Relief and recovery. Owning an account in a financial institution to which government or donors can transfer cash assistance facilitates the provision of aid during disaster relief operations. Immediately after a disaster, some FSPs reschedule or write off loans of clients that have been affected. The FSPs may also provide new emergency loans at very minimal cost and with a longer and/or relaxed repayment policy. Such loans are most effective immediately after a disaster to address financial shocks arising from the loss of employment or savings.

To assist clients with recovery after a disaster, FSPs may offer loans to restore households' assets to the pre-disaster condition. Clients may use such loans to repair damaged assets (e.g., houses and business equipment and infrastructure) and purchase new income-generating assets. During disasters, loan repayments stop or slow down. As a result, some FSPs do not have the capital or liquidity required to meet all their clients' demands for recovery, reconstruction, and rehabilitation. Hence, an emergency liquidity fund for FSPs may be set up to meet their liquidity needs during disasters.

The most suitable option for enhancing women's economic resilience to disasters appears to be precautionary savings and insurance. While credit may be used to finance risk-reducing infrastructure and recovery efforts, the impact on a woman's financial well-being should be carefully considered because credit needs to be repaid along with any interest on a loan. Repaying the debt may impact women's living standards. Repayment difficulties are particularly a concern among low-income women. **Thus, savings and insurance are better suited than loans for women, particularly those in the lower-income bracket**, because most women are used to setting aside small amounts for emergencies and/or specific future needs. Savings and insurance products should, however, be designed to address specific gender barriers that women face. An important consideration when designing financial instruments for women is to use channels and agents that are familiar and accessible to women. Some women prefer to deal with female agents and/or distributors of financial products and services.

Digital technology along with a well-developed payment system, good physical infrastructure, appropriate regulation, and vigorous consumer protection safeguards can facilitate access to and usage of financial products and services. Using digital technology to provide cost-efficient and tailored financial products and services can help strengthen the resilience of people considered vulnerable and prepare them for hazard events. Digital technologies also enable financial institutions to be resilient in times of disasters. Cloud, solar, and satellite technology can play a crucial role in enhancing the resilience of FSPs especially when disasters result in the loss of connectivity (e.g., electricity blackouts, blocked roads, and Internet outages).

Stakeholders need to work together to facilitate and encourage the development, design, and promotion of relevant and appropriate financial products and services that will enhance women's resilience to disasters.

> ▸ **Policy makers and regulators** are urged to (i) adopt gender-specific policies and strategies in national financial inclusion strategies; (ii) establish policies that enable the use of bank accounts and digital financial services for distributing government assistance, particularly during disasters; and (iii) adopt regulations that facilitate women's access to financial services.

> ▸ **Development partners** are requested to support the (i) incorporation of a strategic gender approach in country financial inclusion agendas, (ii) establishment of financial mechanisms that will enable FSPs to meet the financing needs that can enhance women's economic resilience in times of disaster, (iii) conduct of action research on the design of relevant and appropriate financial products for women's economic resilience to disasters, and (iv) conduct of digital and financial literacy programs appropriate for and tailored to women.

> ▸ **FSPs** are encouraged to (i) incorporate the collection of sex-disaggregated data as part of FSPs' regular reporting systems, (ii) provide women access to secure and private savings accounts, (iii) collaborate with organizations that work directly with women as clients and/or as members, (iv) adopt innovative digital technologies to provide access to unserved women and employ alternative forms of credit assessment, and (v) include women staff in designing and developing financial products.

INTRODUCTION

Disasters brought about by natural hazard events and extreme weather conditions resulting from climate change, urbanization, and increasing population densities affect men and women differently. Although natural hazards are gender neutral, differences in the socioeconomic status, ability to access resources, and prevailing social and cultural norms between men and women impact the way they respond to and are affected by disasters. People living at the base of the economic pyramid who have no or limited access to resources are the most vulnerable and least able to cope. Given that the female labor force participation rate is low, women are more likely to live in the poorest households—in 41 out of 75 countries researched, women who are separated, widowed, single mothers, or heads of households without a male partner have an increased risk of living in poverty (Miles and Wiedmaier-Pfister 2018). Women are also more likely than men to be self-employed in the informal economy, with limited discretionary spending due to societal constraints, fluctuating cash flows, fewer assets (due to inheritance customs and restrictions on land and asset ownership), restricted access to the formal sector due to unpaid caring responsibilities, and lower levels of education and literacy (Miles, Wiedmaier-Pfister, and Dankmeyer 2017). With frequent exposure and high vulnerability to shocks, and with less ability to cope and recover from them, women may have limited ability to withstand the impact of disasters resulting in unexpected financial shocks and economic distress. Thus, women are more vulnerable than men to climate change and disasters.

Financial inclusion enhances women's economic resilience to disasters by providing them access to various types of financial products and services. Use of relevant and appropriate financial instruments enables women to prepare for and cope with the effects of disasters. This publication reviews the use and availability of financial instruments that enhance women's resilience to external shocks in disaster-prone and low-income areas.

BUILDING ECONOMIC RESILIENCE TO DISASTER THROUGH FINANCIAL INCLUSION: THE NEXUS

The impact of any disaster is determined by one's exposure and vulnerability to a hazard event and response to disasters is determined by an individual or household's economic resilience and coping skills. The Intergovernmental Panel on Climate Change (IPCC) defined resilience as the ability of a system and its parts to anticipate, absorb, accommodate, or recover from the effects of a hazardous event in a timely and efficient manner by ensuring the preservation, restoration, or improvement of its essential basic structure and functions (IPCC 2012). Erman et al. (2021) stated that economic resilience to disaster is influenced by the (i) level of preparedness, which depends on socioeconomic status, risk perception, education, access to information and media, and previous disaster experience; and (ii) availability of coping mechanisms, which includes access to finance, access to government support, ability to switch income sources, and ability to adapt through migration (Erman et al. 2021). **Economic resilience to disasters is a result of the level of preparedness and coping capacity of an individual or household.** Socioeconomic status and the prevailing social and cultural norms in a society influence women's level of preparedness and ability to cope. For example, the Philippines' *bayanihan*—the spirit of civic unity and cooperation—helps families and individuals cope with the impact of disaster. In the same way, people living in disaster-prone areas are usually more prepared to handle disasters, given their experience, than are people in other areas.

Financial inclusion is a state in which everyone, especially vulnerable people, has effective access to a wide range of financial services (savings, credit, insurance, remittances, and payments).[1] The key dimensions of financial inclusion include (i) access—i.e., that financial service providers (FSPs) are open and within reach by all; (ii) usage—that financial products and services are affordable and adequate, leading to actual usage; (iii) quality—that product attributes match the needs of customers; and (iv) impact—the effect on the livelihoods of customers (AFI 2009). Cutting across these key dimensions is financial literacy. It provides individuals appropriate knowledge and information and relevant tools to make informed financial decisions.

The ability to use relevant financial products and services contributes to risk management in times of adverse shocks. Using financial instruments that are appropriate for risk reduction, risk preparedness, and risk transfer (such as through insurance, for example) increases an individual's or household's level of preparedness for a disaster. The provision of relief and recovery measures using suitable financial products enables individuals and/or households to cope during and after a disaster. The use of digital financial services enables financial institutions to deliver financial products and services in a cost-efficient manner, particularly in times of disasters. Figure 1 shows the link between financial inclusion and building economic resilience.

---

[1] As stated in the National Strategy for Financial Inclusion in the Philippines. The definition is based on several internationally accepted definitions set forth by entities such as the World Bank, the Consultative Group to Assist the Poor, and ADB. https://www.bsp.gov.ph/Pages/InclusiveFinance/NSFI-2022-2028.pdf.

Figure 1: Financial Inclusion, Disaster Preparedness, and Coping Mechanisms

BEFORE A SHOCK			DURING AND AFTER A SHOCK	
Level of Preparedness			**Coping Mechanism**	
Risk Reduction	Risk Preparedness	Risk Transfer	Relief	Recovery
Credit and/or goal-based savings can lead to adoption of risk mitigating technologies (e.g., climate proof houses, climate resilient seeds/seedlings for agriculture) that will lower exposure to or impact of shocks.	Liquid accounts like savings will result in precautionary savings that will prepare them for shocks. Precautionary savings may be used to smooth consumption in the event of a shock.	Insurance transfers the risk to the insurance provider and will provide protection to the insured in the event of a shock. Claim proceeds can be used to rebuild and restore affected assets during a shock.	Having and owning an account in a financial institution will facilitate cash transfers during relief operations. Cash transfers may come either from government or from remittances from family and friends.	Savings and proceeds from insurance claims can be used to rebuild and restore assets and businesses lost due to a shock. Availability of credit will provide the needed capital to restore assets and businesses.
DIGITAL FINANCIAL SERVICES				

Source: Author, based on Moore et al. (2019).

Financial inclusion enhances an individual's or household's level of preparedness for disasters. By using credit, savings, and/or insurance, individuals, households, and businesses can anticipate, prepare, and eventually absorb the negative impacts of a disaster through asset creation, consumption smoothing, and emergency assistance.

(i)   Credit facilities may be used to build houses and business infrastructure that are resilient to disasters. Credit can also be used to provide working and operating capital to finance additional livelihood or expand existing businesses. Surplus from business expansion may be set aside as savings for emergency needs. Hence, access to financial capital can reduce vulnerability through asset creation and increased savings.

(ii)   Precautionary savings help to smooth consumption in times of a financial shock brought about by a disaster. The ability to save cash (instead of buying tangible assets) in a safe institution also reduces the risk of losing assets in the event of a disaster. Thus, the availability of savings products that are tailor-fitted to the needs of women (e.g., small amounts of savings deposited on a frequent basis) will encourage them to set aside a portion of their income for precautionary savings.

(iii) Insurance transfers the risks associated with disasters to the insurance provider. Access to insurance products prior to a disaster therefore emboldens individuals and/or households to invest in activities that generate higher income (e.g., agriculture production) because they know that they can rely on insurance claims when an insured peril occurs that negatively impacts an income-generating activity.

Financial inclusion boosts an individual's or household's ability to cope during and after a disaster. When an individual has an account in a financial institution, cash transfers from government and remittances from family and friends can immediately be facilitated in the event of a disaster. Also, access to credit facilities can provide the resources needed for recovery and reconstruction after a disaster, making it possible to rebuild faster and better (Hallegate et al. 2017). Savings and proceeds from insurance claims may also be used to finance restoration, rebuilding, and recovery-related activities. Immediate access to funds (either through emergency credit or savings) through financial inclusion helps smooth income and consumption right after a disaster.

Digital financial services play an important role in financial inclusion and enhancing economic resilience to disasters. Digital financial services provide a cost-efficient way to reach and provide financial assistance to the most vulnerable people. In the aftermath of a disaster, digital financial services can be used to facilitate and expedite cash transfers from government and/or donor institutions during relief operations.[2] Similarly, digital financial services can also facilitate remittances from family and friends who are not affected by disasters. Transfers facilitated by digital services can then be used to restart and reestablish livelihoods and economic activities following a hazard event. As demonstrated during the coronavirus disease (COVID-19) pandemic, the use of digital financial services greatly facilitated the transfer of cash assistance to those affected by the lockdowns. It also expedited payments for goods and services when direct contact was considered a COVID-19 threat.

The use of technology (e.g., cloud technology) also empowers financial institutions to be resilient in times of disasters. Cloud, solar, and satellite technology can play a crucial role in enhancing the resilience of FSPs especially when disasters result in the loss of connectivity (i.e., no electricity, blocked roads, no Internet).

---

[2]   Cash transfer programming can be used as a form of humanitarian response that provides for basic needs and can be used to protect, establish, or restart livelihoods and economic activities following a hazard event.

WOMEN AND FINANCIAL INCLUSION: A BRIEF OVERVIEW

Financial inclusion is recognized globally as a key enabler in the fight against poverty and in the pursuit of inclusive growth and development. To identify, adopt and implement specific strategies and measures to systematically accelerate financial inclusion, various countries developed and formulated their own national financial inclusion strategy (NFIS).[3]

Studies indicate that countries that have adopted a strategic policy approach to financial inclusion have contributed to an improved state of financial inclusion worldwide. The proportion of adult population reported as unbanked declined from 49% in 2011 to only 31% in 2017. The level of financial inclusion, however, varies across economies. In high-income economies, only 6% of the adult population does not have an account in a financial institution or with a mobile money provider, while 37% of the adult population is still considered unbanked in developing economies (World Bank 2018b).

In Asia, only in India, Malaysia, and Thailand did more than 80% of the adult population own an account in 2017. In contrast, in Cambodia, the Lao People's Democratic Republic, and Myanmar, less than a third of the adult population own an account (Figure 2). India, Malaysia, and Thailand implemented policies that facilitated access to financial services, and they leveraged the use of digital technology to improve access and delivery of financial products and services. The three countries have deliberately adopted policies and installed infrastructure that supports digital financial services. For example, the large leap in account ownership in India is attributed, among other things, to the adoption of Aadhar, a biometric database that provides a unique identity to each Indian citizen (their national identification). Aadhar made the process of opening an account in a financial institution a lot easier, contributing to improved financial inclusion (Iyer 2019). Malaysia's high level of financial inclusion is due in part to policies taking advantage of mobile phones and banking agents to expand access (Martinez 2017). In Thailand, the "fintech" industry was tapped to extend financial services to the underserved and unserved population (Banqin 2020).

Despite the global improvement in account ownership,[4] gender disparities persist, with male account ownership at 72% versus female account ownership at 65%. The 2017 Global Findex reported that the 7-percentage point gender gap has not changed since 2011 among developing economies. Inequality in

---

[3] An NFIS is a comprehensive public document that presents a strategy developed at the national level to systematically accelerate the level of financial inclusion. An NFIS should be developed through a broad consultative process involving, among others, public and private sector stakeholders engaged in financial sector development. Typically, an NFIS will include an analysis of the current status of and constraints on financial inclusion in a country, a measurable financial inclusion goal, how the country proposes to reach this goal and by when, and how it will measure the progress and achievements of the NFIS. (Definition developed by the members of the AFI Financial Inclusion Strategy Peer Learning Group.)

[4] The data on account ownership are from the 2017 Global Findex Database, wherein account ownership means having an account at a financial institution or a mobile money provider. A financial institution may be a bank, a microfinance institution, or a cooperative (World Bank 2018b).

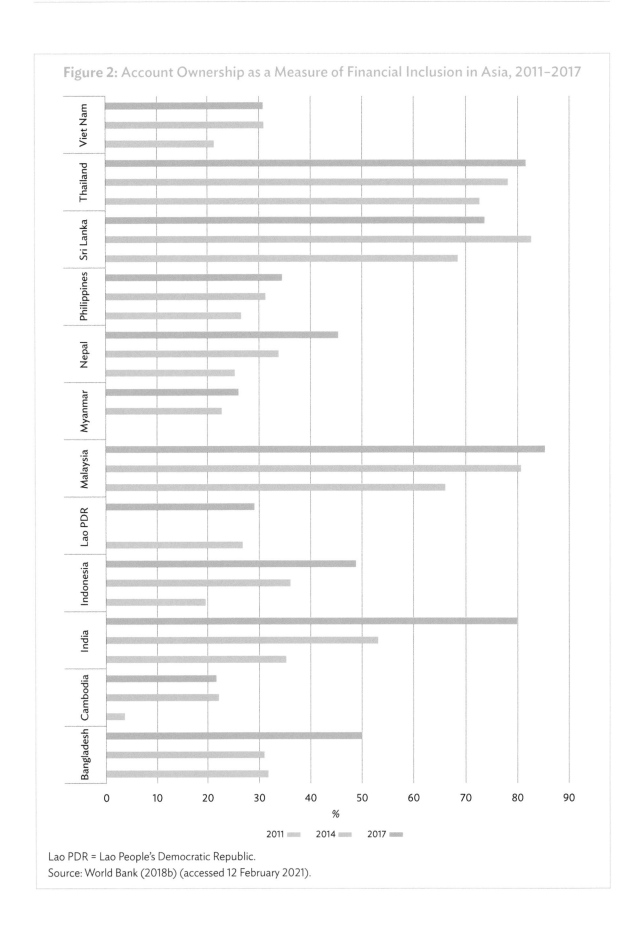

Figure 2: Account Ownership as a Measure of Financial Inclusion in Asia, 2011–2017

Lao PDR = Lao People's Democratic Republic.
Source: World Bank (2018b) (accessed 12 February 2021).

account ownership between the rich and the poor also persists—with a 13-percentage point global gap between households in the richest 60% and those in the poorest 40%.

Access to financial services enables people to improve their management of their financial resources. Enhanced capacities and capabilities to manage financial resources and build their own businesses give women more autonomy. Women's earnings may also be less affected by emergency or disaster-related shocks when they have access to financial products such as savings, credit, and insurance. Thus, use of relevant financial instruments enhances women's abilities to continue their business activities while decreasing disruptions of their earning capacities.

Despite the importance of financial inclusion for women, women still comprise 56% of all unbanked adults and a global gender gap in account ownership persists (World Bank 2018c). The gender gap in financial inclusion is partly explained in the latest edition of the Global Microscope (EIU 2020). The report disclosed that only one-third of the 55 countries covered by the study include a gender approach in their NFIS and almost none has established clear objectives on closing the gender gap (Plata 2021). Only a few governments have taken proactive steps to reduce the gender gap in account ownership; and only one-fourth of the 55 countries collect data on women's use of financial services that could help government identify factors that prevent women from achieving financial inclusion. Some examples of governments that have taken a gender approach in their financial inclusion strategies are as follows (EIU 2019):

(i)   Madagascar is implementing a national financial inclusion project that aims to reach 8,000 new female microfinance customers and provide mobile-based financial education to 30,000 women.

(ii)   Mozambique's financial inclusion strategy sets targets for increasing women's access to savings, credit, and e-money accounts by 2022.

(iii) Pakistan aims to increase the number of women-owned digital accounts to 20 million (although this is still less than one-third of the total number of accounts).

(iv)   A public–private agreement in South Africa aims to increase black women's participation in the financial sector.

Figure 3 shows that, among selected Asian countries, only Indonesia, the Lao People's Democratic Republic, and the Philippines had an account-owning gender gap in favor of women in 2017. In most other countries, the gender gap has favored male account ownership since 2011, as reported in the Global Findex database. In India, Myanmar, and Sri Lanka, the gender gaps narrowed, but it widened notably in Bangladesh.[5] The widening gap in Bangladesh is due to lower participation of women in the workforce, social norms that limit women's mobility and give men most of the financial decision-making power, and the FSPs' lack of products and processes that address the unique needs of women (El Zoghbi 2020).

---

[5]   Two-thirds of men have an account with a financial institution, but less than half of the women do—a 29% gender gap in account ownership.

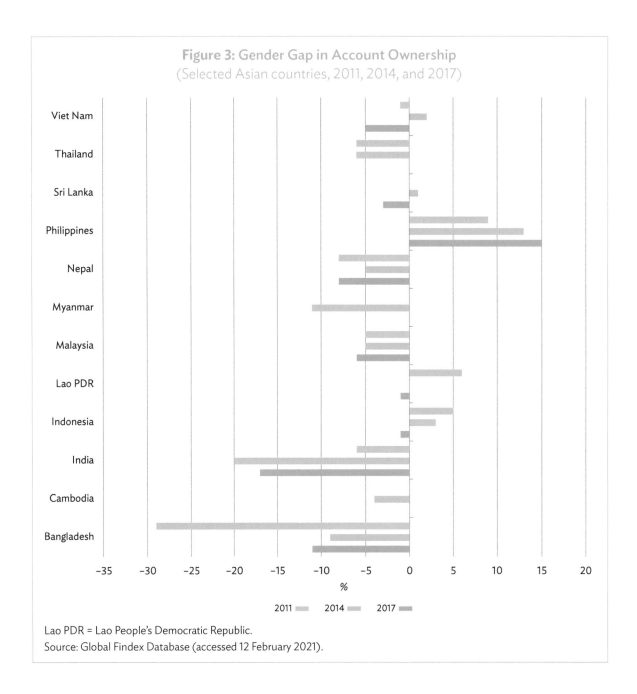

Figure 3: Gender Gap in Account Ownership
(Selected Asian countries, 2011, 2014, and 2017)

Lao PDR = Lao People's Democratic Republic.
Source: Global Findex Database (accessed 12 February 2021).

A "reverse gender gap" situation on account ownership does not necessarily ensure women's access to and usage of financial services. Closer analysis of financial inclusion access and usage reveals more nuanced but important gender gaps. For example, the Asian Development Bank (ADB)-supported nationwide survey of micro, small, and medium-sized enterprises (MSMEs) that focused on women-owned and women-led MSMEs revealed that only 20% of small business owners surveyed reported using any form of loan facility, including invoice finance, overdrafts, letters of credit, and business loans. **Most of the MSMEs surveyed** (most were women-owned or women-led) **were "banked but not yet banking."**

Several factors impede women's access to financial services. Women face multiple barriers to accessing and using financial services. Most of the barriers are a result of sociocultural practices and how women are perceived in society, but some barriers are institutional or merely infrastructure related. The latter include women lacking an identification card (ID) to prove their identity, having insufficient or no traditionally required collateral, facing mobility constraints, and having limited financial literacy (Trivelli 2018). Women also face legal and regulatory barriers that impede their access to financial services. Table 1 shows various demand- and supply-side barriers to financial inclusion.

Table 1: Gender-Based Barriers to Financial Inclusion

Demand-Side Barriers	Supply-Side Barriers	Legal and Regulatory Barriers
Lack of bargaining power within the household	Inappropriate product offerings	Account opening requirements that disadvantage women
Concentrate in lower-paying economic activities	Lack of gender-specific policies and practices for product design and marketing	Barriers to obtaining formal identification
Competing demands on women's time related to unpaid domestic work	Inappropriate distribution channels	Legal barriers to owning and inheriting property and other collateral
Lack of assets for collateral		Lack of gender inclusive gender reporting system
Lack of formal identification		
Reduced mobility due to time constraints or social norms		
Lower rates of cellphone ownership among women, needed to access many digital products		

Source: Holloway, Niazi, and Rouse (2017).

Other barriers to women's account ownership include the following:

(i)   **Cost concerns and perceptions of utility.** A prevailing perception among women is that they do not have a use for an account, or enough income to open an account and maintain it.

(ii)   **Lack of sex-disaggregated data on use of financial products.** In most countries, there is no specific mandate for financial institutions to collect and report sex-disaggregated data that could be used to determine policies relevant to enhancing women's financial inclusion. In addition, there are no institutional processes that regularly capture and analyze sex-disaggregated data to help policy makers, regulators, and FSPs understand the needs, circumstances, and preferences of women (Salman and Nowacka 2020). Sex-disaggregated data can provide financial institutions specific information regarding the viability of investing in women's financial needs. Such data can enable the institutions to appreciate the commercial profitability of women as a client segment.

(iii) **Lack of financial products that cater to women's needs.** Without relevant sex-disaggregated data, FSPs have difficulty designing products that are tailored to the unique needs of women. For example, women have different preferences than men for liquidity and privacy in their choice of savings products depending on their personal and household investment goals and on their bargaining power in the household (Holloway et al. 2017). Further, women prefer to use delivery channels that they are familiar with. Without understanding women's needs and preferences, products, services, and delivery channels may not be appropriate for women.

(iv) **Inability to supply proof of identity as part of "Know Your Customer" requirements.**[6] Many women are unable to comply with this requirement if they belong to excluded segments of the population (i.e., are in the informal economy) that are not issued government IDs.

(v) **Lack of assets acceptable as collateral and lack of credit history.** Banks still prefer assets as acceptable collateral. Women often lack collateral because of their relatively low income and low labor force participation rates.

(vi) **Lack of credit history.** Some credit registries exclude the credit data set of some microfinance nongovernment organizations (NGOs), where most women have credit track record. In this situation, banks are not able to verify and determine the credit history of such women.

(vii) **High costs of dealing with FSPs.** With various demands on women's time arising from their familial and reproductive roles, women may be constrained from engaging with FSPs, particularly if they are located beyond the woman's immediate neighborhood.

(viii) **Low financial literacy rates.** Given their specific roles in the household, women need to understand the importance of financial products and services. Women also need to be informed about and be able to appreciate and understand the processes involved in accessing and using financial products and services.

The foregoing list of demand- and supply-side barriers to financial inclusion highlights the importance of designing appropriate products with features that would help to overcome the barriers.

---

[6] "Know Your Customer standards are designed to protect financial institutions against fraud, corruption, money laundering, and terrorist financing. It involves several steps to establish customer identity; understand the nature of customers' activities and qualify that the source of funds is legitimate; and assess money laundering risks associated with customers." Adapted from SWIFT. Know Your Customer. https://www.swift.com/your-needs/financial-crime-cyber-security/know-your-customer-kyc/meaning-kyc#:~:text=illicit%20criminal%20activities.-,Know%20Your%20Customer%20(KYC)%20standards%20are%20designed%20 to%20protect%20financial,of%20funds%20is%20legitimate%3B%20and.

FINANCIAL INSTRUMENTS TO INCREASE WOMEN'S ECONOMIC RESILIENCE TO DISASTERS

ndividuals and households come up with specific measures and mechanisms to prepare for and cope with shocks and disasters. **Globally, 54% of adults reported that they are able to obtain emergency funds,[7] with a higher proportion reported in high-income economies (73%) than developing economies (50%).** As shown in Figure 4, respondents named savings, money earned from working, and money from family and friends as their major sources of emergency funds.[8]

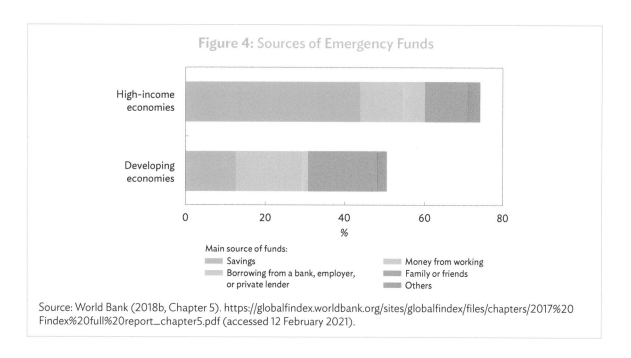

Figure 4: Sources of Emergency Funds

Main source of funds:
- Savings
- Borrowing from a bank, employer, or private lender
- Money from working
- Family or friends
- Others

Source: World Bank (2018b, Chapter 5). https://globalfindex.worldbank.org/sites/globalfindex/files/chapters/2017%20 Findex%20full%20report_chapter5.pdf (accessed 12 February 2021).

Women are often disproportionately more vulnerable to disasters because basic services and assistance are "gender neutral" and do not consider unequal gender roles and discrimination faced by women in many societies. **Gender-responsive financial mechanisms that enable access to financial services reduce women's vulnerability, strengthen their capacity to cope with natural hazard events, and help ensure that they are not left behind when economies rebuild** (AFI 2021).

---

7 The Global Findex Database defined capacity to obtain emergency funds as the ability to access 1/20 of gross domestic income in local currency within a month. The 2017 Global Findex survey asked respondents whether it would be possible to come up with an amount equal to 1/20 of gross national income ... per capita in local currency within the next month. It also asked what their main source of funding would be (World Bank 2018b).

8 World Bank. (2018b): Chapter 5. https://globalfindex.worldbank.org/sites/globalfindex/files/chapters/2017%20Findex%20full%20 report_chapter5.pdf (accessed 12 February 2021).

In times of shock or disaster, capacity of men and women to access emergency funds differ. During emergencies, men use their earnings from work as a source of emergency funds while women tend to rely more on savings and money received from family and friends (Erman et al. 2021). Figure 5 shows that in the Asia and Pacific region, women tend to rely on savings and on money from family and friends as sources of funds in times of emergency. High reliance on informal finance such as family and friends in times of disaster has an impact on women's economic resilience inasmuch as these are not stable and reliable sources of funds during emergency situations. Informal finance may also be difficult to access in times of disaster.

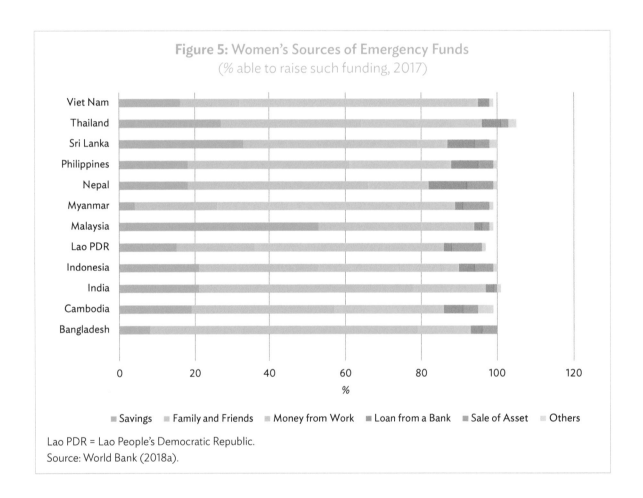

Figure 5: Women's Sources of Emergency Funds
(% able to raise such funding, 2017)

Lao PDR = Lao People's Democratic Republic.
Source: World Bank (2018a).

Risk Reduction

Risk reduction is anything that reduces the chance of losses occurring. Disaster risk reduction is defined as the concept and practice of reducing disaster risks through systematic efforts to analyze and manage the causal factors of disasters, including through reduced exposures to hazards, lessened vulnerability of people and property, wise management of land and the environment, and improved preparedness for adverse events (Government of the Philippines 2010). Disaster risk reduction is the process of protecting the livelihoods and assets of individuals and communities from the impact of hazards. This includes any of the following: building necessary infrastructure (e.g., retaining walls, dams, embankments) to reduce risk; natural resource management; agricultural interventions (e.g., use of resistant crop varieties); and safe shelter/housing (Concern Worldwide 2015). Income diversification is also a risk-reducing activity. With multiple sources of income, risks, and vulnerabilities to the impacts of extreme weather events and climate change may be reduced.

Because disasters such as floods, cyclones, and typhoons occur at a specified time during the year in Bangladesh, some microfinance institutions (MFIs) designed their loan products to help clients minimize exposure to losses resulting from disasters. For example, instead of a weekly or equal amount of loan repayments over a 12-month period, MFIs working in or near a river valley adjusted their loan repayment schedules so that no principal payments are due during the flood season. With this type of loan product, clients, mostly women, are given greater security when taking a loan because the probability of loan default is reduced if a severe flood temporarily prevents them from earning income (Brown and Nagarajan 2000). MFIs also offered loans specifically designed to cope with floods. Using these loans, clients were encouraged to purchase assets that could help them reduce their losses, e.g., construction of resilient, stronger, and more protected houses (Ullah and Khan 2017).

MFIs in Bangladesh also provide pre-disaster loans. The Bangladesh Bank, through its network of government-sponsored MFIs, provides long-term (10-year) loans for clients in flood-prone areas to construct highly flood-resistant cement and tin houses (Brown and Nagarajan 2000). Pre-disaster asset purchased loan products are also offered. These include (i) loans to purchase a small boat that is used for productive purposes during normal times with a condition that during a disaster, the boat will be used to transport fellow community members and their belongings to higher ground; and (ii) a community loan product, which is a loan by all the clients of an MFI and is used to construct a community flood shelter area.[9]

The key features that need to be considered when designing loan products that will reduce disaster risks women face are (i) financing business expansion and/or livelihoods that will enable women to earn additional income and generate surplus for precautionary savings in times of disaster, (ii) financing infrastructure that is climate adaptable using the infrastructure and/or equipment as collateral, (iii) using personal property as acceptable collateral for loan products that reduce disaster risks, and (iv) providing loan repayment terms that anticipate the occurrence of weather hazards to minimize default losses.

Loan products that finance infrastructure, equipment, or activities that will reduce the impact of disaster risk for women should be designed to address specific gender barriers. Table 2 identifies the important features that need to be considered in credit product design, the specific gender barrier addressed, and the benefits and challenges associated with the specific financial products for women.

---

[9]   During normal times, the shelter areas may also serve as sources of income-generating opportunities. Under this program, the MFI uses a portion of clients' accumulated savings to (i) buy a section of land near the community; (ii) employ clients to raise the land above the level of flood waters; and (iii) build income-generating infrastructure on the land (fish ponds, agriculture) in addition to shelters, a tube well, and a sanitary latrine. During floods, this land serves as a safety zone where the community can take their families, livestock, and other household assets (Brown and Nagarajan 2000).

Table 2: Features to Consider in Designing Credit Products for Women

Financial Instrument	Benefits	Potential Challenges	Important Features to Consider in Product Design	Gender Barrier Addressed by the Product Feature
Risk Reduction				
Credit to finance risk-reducing infrastructure (e.g., houses, business infrastructure) Credit to finance economic activity for income diversification	Funds for expenditures related to putting up infrastructure that is climate resilient Funds to set up, expand, or recalibrate businesses to make them climate resilient and/or disaster proof Funds to engage in diversified economic activities	Lack of credit track record with financial service providers Lack of sufficient income to make the necessary repayments that could lead to over indebtedness	Use of alternative forms of collateral such as assets to be financed Simple documentary requirements Repayment terms based on cash flow of women and impact of disasters on cash flow Quick disbursal with option to release to women's e-wallet	Lack of assets for collateral and lack of credit track record Lack of formal identification Inability to supply proof of identify for the Know Your Customer requirement Inappropriate product design for women's needs Reduced mobility due to time constraints or social norms

Source: Asian Development Bank.

Risk Preparedness

Savings can be used to prepare for disasters. They are considered a major source of emergency funds for households when a shock occurs. Precautionary savings are funds set aside prior to a disaster or shock and serve as a form of self-insurance. These funds are used to smooth consumption and manage responses in the event of a shock.

The 2017 Global Findex Report reported that savings are one of the top three sources of funds during emergencies. Savings are also the most preferred financial product among women, who tend to be more risk averse than men. Instead of relying on borrowed capital, women prefer to use savings during emergencies and for business expansion or investments (Buvic and O'Donnell 2016). Women's World Banking research indicates that low-income women can save about 10%–15% of their net monthly income, but this rarely covers more than the basic household emergencies and minor health-related costs (Miles, Wiedmaier-Pfister, and Dankmeyer 2017).

Some savings products designed for women consider their specific needs and preferences. Women prefer commitment savings products that restrict access to funds until a specific goal is reached because such products protect their savings from the demands of friends and family members. In the Philippines, Green Bank developed a commitment savings product known as the Save, Earn, Enjoy Deposit account for women (Ashraf, Karlan, and Yin 2006). It is a savings account with a defined goal (either a date goal or an amount goal). Savings withdrawals are not allowed until the set goal is reached. This type of savings led to increased expenditures for female-oriented and durable goods. To increase women's resilience to disasters, a commitment savings product that sets specific goals such as building climate-resilient houses and/or establishing climate-resilient businesses may be offered to women. Commitment-savings goals may include using the money in the event of disaster to smooth consumption.

No-fee savings accounts are designed to meet the needs of the unserved and underserved, particularly low-income women. In the Philippines, the Bangko Sentral ng Pilipinas (the central bank) issued a regulation in 2018 allowing banks to open basic deposit accounts. A basic deposit account is an interest- or non-interest-bearing account that may be used to receive and make payments and provide a facility to store value and save money. It has simplified Know Your Customer requirements, requires small amounts for account opening, and has no dormancy charges or requirements for maintaining a balance. Given these simple and basic features, banks are given the liberty to customize product offerings based on the needs of the identified market.[10] As of March 2021, there were about 7 million basic deposit accounts in the Philippines (Business World 2021). No fee savings accounts are also offered in Chile, Kenya, and Nepal.[11] No frills savings accounts were also offered in Brazil and India (AFI 2021b). These were offered to low-income individuals and were used to receive cash transfers from government particularly during disasters. A no-fee savings account with a small amount required for opening the account is tailored to the needs and circumstances of low-income women, who may not have sufficient money to open a regular savings account.

In Indonesia and Tanzania, interest-bearing mobile savings products are offered with training sessions on business skills and on the use of the mobile savings platform. Women who availed of the training were able to save more than women who did not receive any training. Business skills training improved women's record-keeping activities and enhanced their financial planning capacities. The training also enabled women to operate a second business, which improved their profit-making capacities.[12] Improved business skills led to increased income opportunities, so that women could be better prepared for disasters or calamities.

Given the time and travel restrictions arising from women's traditional household and care responsibilities, the use of agents conveniently located near women's places of residence is an important innovation some FSPs employ to provide women with access to financial services. For example, the NBS Bank in Malawi offers the Pafupi ("close to you") savings account. Pafupi allows small deposits, has no monthly fees, and operates through agents in conveniently located rural shops (Women's World Banking 2018). Similarly, Diamond Bank in Nigeria offers BETA, a simple and affordable savings account that offers home service through a network of mobile agents known as "BETA Friends." It targets self-employed men and women who want to save frequently. The account can be opened in 5 minutes and has no minimum balance and no fees. The BETA Friends visit a customer's business to open accounts and handle transactions, including deposits and withdrawals, using a mobile phone application.

---

[10] Bangko Sentral ng Pilipinas. 213 Basic Deposit Account. https://morb.bsp.gov.ph/213-basic-deposit-account/.

[11] Global Partnership for Financial Inclusion. G20 Financial Inclusion Indicators. https://datatopics.worldbank.org/g20fidata/.

[12] Center for Global Development. (n.d.). *Mindful Saving: Exploring the Power of Savings for Women* (Factsheet). https://www.cgdev.org/sites/default/files/mindful-saving-exploring-power-savings-women-factsheet.pdf.

In the Philippines, the Bangko Sentral ng Pilipinas allowed the use of cash agents[13] that can accept and disburse cash on behalf of a bank, perform Know Your Customer procedures, and collect and forward loan and account opening applications. Cash agents are also allowed to sell and service insurance as allowed by the Insurance Commission. As highly accessible outlets, cash agents provide a distinct service to clients, women.

Savings through informal and semi-formal groups are another important mechanism that can prepare women for disasters. Savings groups such as village savings and loan associations, self-help groups, and rotating credit and savings associations mostly comprise women. Members of these groups are socially bound and committed to save, thus enhancing a savings culture among members. Because the resources of these savings groups are limited to the total amount of savings of all the members, savings groups tend to be less liquid although more flexible than the savings products offered by formal financial institutions. In view of this, informal savings groups may not be able to address the small but common shocks that may be prevalent in a community. However, savings groups provide social support to their members during disturbances such as weather- and climate-related shocks. An example of this is the Savings for Change program in Mali shown in the box.

Box: The Saving for Change Program in Mali

The Saving for Change Program in Mali provides assistance to women organizing themselves into simple savings and credit groups. The program aims to address the needs of women who are not reached by formal financial service providers or traditional rotating credit and savings associations. About 20 women voluntarily form a group that elects officers, establishes rules, and meets weekly to collect savings from each member. At meetings, each woman deposits a previously determined amount into a communal pool, which grows in aggregate size each time the group meets. When a member needs a loan, she asks the group for the desired amount; the group collectively discusses whether, how, and to whom to disperse the funds. Loans must be repaid with interest, at a rate set by the members, and the interest collected is added to the communal pool of funds. Saving for Change introduced a novel oral accounting system which helps each woman manage her debts and savings totals.

Source: Arnold (2021).

Considering that women are generally risk-averse, savings products are preferred and deemed important for strengthening the economic resilience of women. Precautionary savings are considered low-cost sources of liquidity accumulated over time by regularly setting aside small amounts. Table 3 summarizes the features that should be considered when designing precautionary savings. The table identifies specific gender barriers addressed by specific design features, and the associated benefits and challenges. For example, women prefer commitment savings that restrict access to funds until specific goals are met because that protects their funds from the demands of family for unplanned expenses.

---

[13]   The cash agents are typically cash-rich, third-party entities with many outlets that conduct regular business in fixed locations, such as convenience stores, pharmacies, and other highly accessible retail outlets. Bangko Sentral ng Pilipinas. https://www.bsp.gov.ph/Pages/InclusiveFinance/InitiativesAndRegulation.aspx.

Table 3: Features of Precautionary Savings Plans for Risk Preparedness

Financial Instrument	Benefits	Challenges	Important Features to Consider in Product Design	Gender Barrier Addressed by the Product Feature
Precautionary savings with commitment goals	Low cost and easily accessed in times of disaster (as the commitment goal) Accumulated over a period of time Limits debt Provides a cushion against unexpected losses due to disaster Money is kept safe and cannot be easily accessed for purposes other than stated Women are familiar with setting aside funds for emergency purposes	Inadequate income to set aside for savings Inability to comply with the requirements for opening a savings account Cannot be withdrawn when stated goal is not met	Tied to a specific commitment goal, i.e., withdrawable only when the goal is met Small amount and simple documentary requirements for account opening For digital savings, cash-in, cash-out modes should be available in channels that are familiar and located close to women (e.g., convenience stores, rural agents) Inclusion of training program for women savers No fees and charges	Lack of bargaining power within the household Inability to fend off demands from family members Concentration in lower-paying economic activities Lack of formal identification Competing demands on women's time Reduced mobility due to time constraints or social norms Low utility perception of a savings account Concentration in lower-paying economic activities Cost concerns

Source: Asian Development Bank.

Risk Transfer

Unlike idiosyncratic risks (i.e., risk events limited to an individual and/or household), which may be addressed by precautionary savings, "covariant" or catastrophic risks brought about by natural hazard events such as earthquakes, floods, or typhoons are best addressed through risk transfer or insurance, which transfers the risks of a specific peril to the insurance service provider.[14] Through risk transfer, insurance provides rapid access to post-disaster liquidity offering protection to individuals, households, and/or their livelihoods. As such, insurance can reduce the impact of disasters on clients while increasing their ability to recover and repay loans (Pantojan 2002). Insurance builds financial resilience in the event of an economic shock brought about by a disaster.

---

[14] "Covariant" in this sense is an insurance term referring to a risk affecting many people in the same area at the same time. https://www.openriskmanual.org/wiki/Covariant_Risk.

Insurance products that address disaster risks may be provided directly or indirectly. Indirect insurance is offered at the macro and meso levels while direct insurance products are provided at the micro level. With indirect insurance, the insured/policyholder is either a public or a private institution that buys insurance on behalf of specific beneficiaries. The insured pays the premium; receives the payout in the event of a claim; and determines how the claim benefits will be distributed to the beneficiaries, which is usually based on a prespecified protocol of action (Miles and Wiedmaier-Pfister 2018). With direct insurance, the insured pays the premium and receives the claim benefits when an insured peril occurs.

Indirect macro insurance. In recent years, several countries have designed and implemented sovereign catastrophe risk pools. Under these arrangements, the policyholder is either the national or the local government. These risk pools provide benefits and services to vulnerable populations as part of disaster contingency planning and social protection measures. Payouts are spent for immediate response while humanitarian aid is being mobilized, maintenance of government services and cash flows, funding of pre- and post-disaster operations, and targeted assistance.

A few sovereign risk pools have incorporated some gender considerations in their design and implementation. For example, the Kenya Hunger and Safety Net Program, which is a risk pool implemented by the Government of Kenya, developed delivery channels that enhance the program's accessibility for women. The program has partnered with women's groups and used digital distribution channels to overcome women's mobility constraints. The program has also developed a targeting system that takes into account women's vulnerabilities.

The Caribbean Catastrophe Risk Insurance Facility and the Pacific Catastrophe Risk Assessment and Financing Initiative incorporated some limited gender consideration into their approach. The Caribbean Catastrophe Risk Insurance Facility collects gender impact data and the Pacific Catastrophe Risk Assessment and Financing Initiative collects sex-disaggregated data on beneficiaries and payout audits. Neither facility, however, considered specific gender issues and concerns during their design phases. A study that reviewed the programs notes "There is no evidence that to date any sovereign Climate Risk Insurance scheme has been designed or implemented with an explicit gender equality or women focused target. ...a full assessment is required to verify the extent to which gender considerations are integrated into each of the emergency plans in place by the states signatories of macro policies" (Miles and Wiedmaier-Pfister 2018: p. 23).

A more recent program is the Southeast Asia Disaster Risk Insurance Facility, which serves as a platform for ASEAN+3 countries[15] to access disaster risk financing solutions and increase financial resilience to climate and disaster risks. However, the facility is yet to establish specific measures for incorporating gender considerations.[16]

Gender considerations may be incorporated in the design of indirect macro types of insurance (Miles and Wiedmaier-Pfister 2018). Government may design payouts to directly benefit and provide cover to women who have a lower level of assets and higher vulnerability to disaster risks.

---

[15] Association of Southeast Asian Nations includes Brunei Darussalam, Cambodia, Indonesia, the Lao People's Democratic Republic, Malaysia, Myanmar, the Philippines, Singapore, Thailand, and Viet Nam. ASEAN+3 includes the foregoing plus Japan, the People's Republic of China, and the Republic of Korea.

[16] Southeast Asia Disaster Risk Insurance Facility. www.seadrif.org.

Indirect meso insurance. With indirect meso insurance, groups or institutions that directly work with and for women (e.g., MFIs, NGOs assisting women, and cooperatives with women membership) are the policyholders. An example is the African and Asian Resilience in Disaster Insurance Scheme, which provides insurance to Vision Fund, an MFI whose clientele is about 80% female. In the same way, Afat Vimo, a disaster microinsurance scheme offered by the All India Disaster Mitigation Institute for small and informal business units, partners with the Society for Women Action Development, which is committed to the development and welfare of poor women (Miles and Wiedmaier-Pfister 2018).

In Indonesia, the state reinsurer PT Reasuransi MAIPARK offers earthquake index insurance that provides portfolio-level cover to FSPs such as MFIs. It covers loan portfolio losses due to earthquakes and protects MFIs and/or FSPs from liquidity crises after an earthquake. The insurance payout is triggered by a predetermined earthquake parameter (magnitude and intensity) announced by an independent third party (ADB 2020). This approach has supported women, as the majority of the MFI clients are women.

In Bangladesh, a meso-level flood index insurance program was launched in 2013 in partnership with Manab Mukti Shangstha, a local NGO. The program covered 1,661 poor households with insurance written by Pragati Insurance Limited and reinsurance protection provided by Swiss Re. Oxfam supported and subsidized the program. The policyholder is Manab Mukti Shangstha. Under the scheme, households were given payouts when the water level trigger is breached in an area covered by the insurance (German International Cooperation 2020; International Center for Climate Change and Development 2014). The target participants included "hardcore poor," poor, widows, divorced, and disabled people.

Direct micro insurance. This type of insurance is offered directly to individuals through distribution channels such as MFIs and cooperatives. Women comprise a large majority of the membership of these organizations. The policyholders are directly insured with an insurance provider. In the Philippines, nonlife microinsurance products are offered by 11 nonlife insurance companies, and include coverage for property, livelihood, crops, and calamities (de Vera 2021). This insurance product played a major role in the recovery from Typhoon Haiyan (Swiderek and Wipf 2015). Approximately 111,461 microinsurance claims amounting to $12 million were paid to insured low-income families. The funds were used to rebuild houses and restart livelihood activities (AFI 2021).

In the Philippines, Center for Agriculture and Rural Development (CARD) Pioneer Microinsurance Inc. (CPMI) provides a credit-bundled product that offers flood and typhoon benefits as well as death benefits. The CPMI also offers an indemnity-based micro agriculture insurance product to its partner MFIs and rural banks that serve as distribution channels (German International Cooperation 2020). Most of the clients of their MFI–NGO partners are women. The CPMI also provides for the nonlife insurance needs of CARD Mutual Benefit Associations, which provide life insurance to members of CARD MFIs.[17]

---

[17] To meet the insurance needs of the CARD Bank and CARD MFI–NGO clients, who are mostly women, the CARD Mutual Benefit Association was established. In the Philippines, such associations are only allowed to provide life insurance products to its members. CPMI provides the nonlife insurance needs of its members.

Weather-based index or area yield-based index insurance is also offered to insure farmers (including women farmers) against disasters. Index-based insurance is provided through contracts written against specific perils or events such as drought, flood, and rainfall. Payouts are based on an index such as rainfall or wind speed, measured at a local weather station or satellite rather than by damage caused by weather. When actual rainfall and/or windspeed measurements breach the index, payouts are given to the insured. Area-based yield indexes use yield to determine payouts. Payments are made if the yield for a specific farm falls below a predefined limit, which is usually based on the average yield for a specific locality over a period of time. Bangladesh, India, and the Philippines have pilot tested weather-based and area-yield-based insurance products.

(i)   In Bangladesh, the International Finance Corporation in partnership with the Green Delta Insurance Company and the Global Index Insurance Facility developed an index-based insurance product that addresses perils such as drought, excess rain, heat waves, and cold spells. Instead of selling directly to farmers. Green Delta tied up with various actors in the value chain (e.g., FSPs, input suppliers, and contract farming enterprises) that provide support to farmers (German International Cooperation 2020). While there are women farmers in Bangladesh, it is not clear whether women were able to access the index-based insurance product.

(ii)   In India, the Self-Employed Women's Association provided an integrated package of financial services to its members to reduce their vulnerability to negative shocks. Recognizing that its members cannot afford to interact with multiple companies and manage multiple premiums and policies, the Self-Employed Women's Association designed an insurance product that provides life, health, and asset insurance. The insurance product is offered as a combination of savings and credit through Swashrayee Mandals. Members save for the insurance premium through small monthly installments in their Swashrayee Mandals. When a policy is renewed, premium payments are deducted from the savings accumulated. If the total amount of savings is not sufficient, the balance is treated as a loan from the Mandal and is amortized by the member. The use of interlinked transactions facilitates collection of information, minimizes enforcement problems, and provides poor women access to insurance (Patel and Nanavaty n.d.).

(iii) AXA Philippines provides an insurance product for MSMEs. The product offers business interruption coverage against typhoons and floods. The Department of Trade and Industry provides support by educating MSMEs about disaster and business continuity planning. However, very few MSMEs have availed of the product. A majority of MSMEs in the Philippines are owned and headed by women. Thus, this product caters to the need to enhance women's economic resilience (German International Cooperation 2020).

Table 4 shows the types of risk transfer financial instruments that suit women's unique situations. Important features to consider when designing direct insurance products for women are premium affordability and the use of distribution channels that are familiar to women. These features are needed to address women's income, mobility, and time constraints. The table also identifies the benefits and challenges associated with the instrument. Meanwhile, indirect meso insurance (using organizations that meet the risk protection needs of women) is a good start to orient women on the important benefits of insurance.

Table 4: Risk Transfer Instruments Suitable for Women

Type of Insurance	Benefits	Challenges	Important Features to Consider in Product Design	Gender Barrier Addressed by the Product Feature
Direct insurance	Protects income from unexpected losses caused by disasters Protects savings for intended use and future plans (e.g., for education or business expansion) Protects the family in the event of financial losses resulting from unexpected shocks, e.g., disasters	Lack of understanding of insurance and how it works Insurance is considered a wasteful expense when the covered peril does not happen Premium payment may be high particularly in disaster-prone areas	Affordability of premiums Use of distribution channels that are familiar and accessible to women Bundling insurance with other financial products such as savings and/or credit	Concentration in lower-paying economic activities Reduced mobility due to time constraints Distribution channels that are inappropriate or not known to women Competing demands for women's time Low income of women
Indirect Meso Insurance	More efficient for insurance companies because they deal with one entity Women's organization or meso-level institutions are more familiar with the needs of their members	Risk that claims may not be given to the women member In the case of microfinance institutions as meso organizations, benefits may accrue only to the microfinance institution as protection for its loan portfolio	Member-based women's organization should be consulted in the product design Specific arrangements as to how the benefits will be distributed to women members should be clearly laid out	Reduced mobility due to time constraints Inappropriate distribution channels that are not known to women

Source: Asian Development Bank.

Relief and Recovery

Relief. Government and donor organizations (both public and private) provide relief to ensure that households and individuals can survive until disaster has passed. Aside from distributing relief goods, some organization distribute cash as humanitarian aid. Owning an account in a financial institution to which government or donors can transfer cash assistance facilitates the provision of aid during disaster relief operations.

When a disaster occurs, some FSPs reschedule or write off loans of clients that have been affected by disasters. They also provide new emergency loans at a very minimal cost and with a longer and relaxed repayment policy (Ullah and Khan 2017). Such loans are mostly effective in the short term immediately after a disaster and are used to address financial shocks arising from the loss of employment or savings.

In Bangladesh, quick disbursing loans have helped clients survive the immediate aftershocks of a disaster. Such loans are for small amounts and carry short maturity periods. MFIs in Bangladesh provided disaster relief loans as replacement income sources for affected clients, mostly women, to meet their basic consumption needs until they are again able to engage in income-generating activities. The size, terms, and conditions of such loans vary depending on the MFI's available funds. The loans are made available to clients right after a disaster occurs (e.g., or when the floodwaters are still high in the case of a flood), and repayments start as soon as the client enters into a post-disaster reconstruction phase (Brown and Nagarajan 2000).

MFIs in Bangladesh also provide advances against the savings of their clients during emergency situations (Tighe n.d.). The advances are used to cover immediate relief-related expenditures after a disaster.

Recovery. Recovery involves the rehabilitation and reconstruction of structures that were adversely affected by the disaster. This includes assistance to repair dwellings and community facilities, restore businesses and their accompanying infrastructure, and facilitate the revival of economic activities (Ullah and Khan 2017).

To assist clients in post-disaster recovery, financial institutions provide loans to restore households' assets to the pre-disaster condition. In Bangladesh, members of microfinance groups were given loans to repair damaged assets, purchase new income-generating assets, and repair houses (Tighe n.d.). The loan terms and conditions vary among MFIs depending on the (i) income-generating potential of the asset to be financed (e.g., loans for household repairs might have more lenient terms than loans for livelihood or economic activities that will immediately begin producing additional income); and (ii) a household's capacity to repay given the damage done by the disaster. For example, the Bangladesh Rural Advancement Committee (BRAC) offers a reconstruction loan program for income-generating assets. Loan disbursal is given in-kind as BRAC provides borrowers with replacement assets such as seeds, poultry, livestock, and saplings (Brown and Nagarajan 2000; Pantojan 2002). Several MFIs in Bangladesh also provided short-term support services such as donations for health treatment, purchase of essential household commodities, and humanitarian services.

Housing improvement loans are provided to clients especially after a disaster. For example, CARD in the Philippines introduced a specialized housing scheme in collaboration with a subsidiary agency. The housing loan was made available to clients who live in vulnerable areas and are affected by disasters (Ullah and Khan 2017). CARD's clients are mostly women.

Some FSPs also provide assistance in cash or in kind, as was done by the Grameen Bank during the 1998 flood in Bangladesh. The Grameen Bank also allowed affected clients to withdraw money from their Group Fund Savings. Grameen Bank also provided new loans to borrowers who had paid off at least half of their outstanding loans. The loan amount provided is equivalent to what had been repaid prior to the disaster (Pantojan 2002). Similarly, the Association for Social Advancement in Bangladesh offers a permanent disaster loan product that carries no interest and must be paid within 2 years through 100 equal weekly installments (Pantojan 2002).

Pallli Karma Sahayak Foundation (PKSF) in Bangladesh established the Disaster Management Program that provides a fund called SAHOS. Under this fund, households that are affected by disasters are given quick access to emergency credit. The PKSF also waives the entire loan amount of a borrower if the household's income earner is killed or disabled due to a disaster (Dhakal, Simkhada, and Ozaki 2019).

During disasters, loan repayments stop or slow down. Because of this, some FSPs may not have the capital or liquidity required to meet all their clients' demands for recovery, reconstruction, and rehabilitation. In view of this, several mechanisms have been instituted and/or implemented to assist FSPs (mostly MFIs).

(i)   **The Emergency Liquidity Fund (ELF) for MFIs in Latin America.**[18] The ELF was set up using donor capital and is used to provide liquidity in case of a hazard or an external shock. Based on a set of criteria, ELF funds are channeled to well-managed and efficient MFIs that then quickly lend the funds to clients that were affected by hazard events. Qualified MFIs are allowed to take out a 6-month loan to cover their short-term liquidity needs. The MFI may opt to extend the loan up to 2 years but with each extension, the interest charges increase to discourage long-term dependency on the ELF funds.

(ii) **Disaster Loan Fund** (Pantojan 2002). This fund was set up in Bangladesh by CARE, an international NGO, after the 1998 flood. The fund was partly capitalized by recycling aid funds used in the Post Flood Rehabilitation Assistance project with additional donor funds. Funds were released to CARE, which then transferred them to eligible and qualified MFIs. Qualified MFIs then transferred the funds to qualified clients as emergency loans after a disaster.

(iii) **Disaster Management Fund** (Pantojan 2002). This fund was set up by the PKSF, an apex foundation established by the Bangladesh government, to provide low-cost funds to poverty-oriented MFIs. As an apex institution, the PKSF receives funds from donors. After the 1998 floods, the PKSF rapidly disbursed loans to its partner organizations. It also set aside grant contributions to set up a more permanent disaster management fund. The fund can be accessed by partner organizations in the event of localized disasters affecting a small number of clients, and when a national disaster is officially declared.

(iv) **Fiji's Disaster Rehabilitation and Containment Facility** (AFI 2021). The facility was initiated in 2009 with funding from the Reserve Bank of Fiji. The facility provides support for businesses to restart and rebuild following serious disasters (e.g., floods and tropical cyclones). The facility was expanded to support home repairs and rebuild and restart businesses affected by disasters. The funds are lent to lenders approved by the Reserve Bank of Fiji, which in turn lend to homeowners and business owners, including women-owned enterprises.

---

[18] Administratively, the ELF is one fund among several managed by Omtrix, a financial consulting firm based in San Jose, Costa Rica. Most of the ELF came from the Swiss State Secretariat for Economic Affairs and the Inter-American Development Bank, although several nonprofit and corporate investors also contributed. The ELF itself generates little income—members pay only a small one-time membership fee—and because lending is contingent on the occurrence of a hazard, interest income is unreliable. The ELF is grouped with other funds, which spreads overhead costs and contributes to income from equity management and makes the daily maintenance of the fund feasible (Jacobsen, Marshak, and Griffith 2009: p. 16).

Credit facilitates recovery. The foregoing shows that credit to smooth consumption and finance reconstruction and rebuilding efforts after a disaster is important to facilitate recovery. However, as identified in Table 5, credit may also lead to indebtedness particularly among people with insufficient income to repay loans after a disaster. Women's concentration in low-paying economic activity may pose a barrier to their use of credit to enhance economic resilience during disasters. The table identifies the relevant financial instrument for relief and recovery as well as important design features that should be considered to address specific gender barriers.

Table 5: Financial Products for Relief and Recovery

Financial Instrument	Benefits	Challenges	Important Features to Consider in Product Design	Gender Barrier Addressed by the Product Feature
Account Ownership in a Financial Institution	Useful for receiving cash assistance and/or donations in the event of disaster	Lack of documentary requirements for account opening Lack of knowledge on how to open an account	Simple documentary requirements and small amount for account opening	Concentration in lower-paying economic activities Inability to supply proof of identify for Know Your Customer requirement
Advances Against Savings	Limits debt	May not be enough to cover the financial requirements in the event of disaster	Inclusion of literacy and education program on use of savings for emergency purposes	Low perception of utility of a savings account
Credit to finance rebuilding and recovery of businesses and/or enterprises **Credit** to refinance existing loans	Provide funds for rebuilding of properties and recovery of business Provide funds for major expenses that can be paid over a period of time at affordable regular installments	Not easily accessed particularly for people without a credit track record Inability to repay can result in overindebtedness particularly an issue in times of disasters Interest on loans can be costly	Simple documentary requirements Quick loan disbursal Repayment terms designed in consideration of cash flow after a disaster	Inability to supply required proof of identity Reduced mobility due to time constraints or social norms
Credit to smooth consumption right after a disaster. This may come in the form of refinancing existing loans.	Provide immediate source of liquidity during disasters	Financing institutions may not have sufficient liquidity	Low cost and short-term maturity Quick loan disbursal	Low income

Source: Asian Development Bank.

Emerging Financial Instruments
for Women's Economic Resilience

It is important for women to have sufficient access to liquid assets to withstand shocks from disasters and/or life events without serious impact on their income and assets. As shown in the examples of financial instruments, access to liquidity in times of disasters may come from precautionary savings; insurance proceeds; cash assistance from government, donors, family, and/or friends; and loans from FSPs or family and friends.

The following financial instruments help enhance women's level of preparedness for disasters: (i) **loans** that finance infrastructure (e.g., disaster-resilient housing) and economic activities that reduce the risks women face in the event of disasters (e.g., disaster-resilient infrastructure); (ii) **precautionary savings** that can smooth consumption in the face of a shock; and (iii) **insurance**, which protects women's income and assets (i.e., physical and financial assets such as savings) in the event of disaster.

Financial instruments that can bolster the coping capacity of women during and after a disaster are (i) **account ownership** in a financial institution for the efficient transfer of cash assistance to during relief operations, (ii) **easily withdrawable savings** to smooth consumption, (iii) **short-term quick disbursing loans** to smooth consumption immediately after a disaster, and (iv) **loans** to finance women's rebuilding and recovery efforts after a disaster.

The most suitable option for enhancing women's economic resilience to disasters appears to be precautionary savings and insurance. While credit may be used to finance risk-reducing infrastructure and recovery efforts, its impact on a woman's financial well-being should be carefully considered because credit needs to be repaid along with any interest on a loan. Women's living standards may be impacted while they are repaying debt. Repayment difficulties are particularly a concern among low-income women. **Thus, savings and insurance are suited for women, particularly those in the lower-income segment of the population,** because most of such women are used to setting aside small amounts for emergencies and/or specific future need. However, these products should be designed to address specific gender barriers that women face.

An important feature of financial instruments for women is the use of channels and agents that are familiar and accessible to women. In some cases, women prefer to deal with female agents and/or distributors of financial products and services.

LEVERAGING DIGITAL TECHNOLOGY

wo-thirds of unbanked adults have a mobile phone and about 29% of adults use the Internet to pay bills or shop online (World Bank 2018b). This was widely demonstrated during the lockdowns brought about by the COVID-19 pandemic. The COVID-19 threat highlighted the importance of digitization and underscored the convenience of using mobile and e-money for online payments and transfers. Economies have recognized the opportunities that digital technology provides to overcome barriers that unbanked adults face and advance financial inclusion. Digital technology along with a well-developed payment system, good physical infrastructure, appropriate regulation, and vigorous consumer protection safeguards can facilitate access to and usage of financial products and services. Using digital technologies to provide cost-efficient and tailored financial products and services can help to strengthen the resilience of people considered vulnerable and to prepare them for hazard events (AFI 2021).

Using digital technology to provide financial products and services addresses some of the barriers to women's financial inclusion. Following are some examples:

(i) **Digital financial services address women's time constraints.** Experience from a 5-month relief program in Niger that switched the monthly payment of government social benefits from cash to mobile phones saved the recipients 20 hours on average in travel and waiting time to obtain their payments (World Bank 2018b). This is especially beneficial for women considering that transaction costs are a significant barrier for women to use financial products and services. The time saved can be used to care for the household or engage in income-generating activities, which could further improve women's resilience to disasters (Haworth et al. 2016).

(ii) **Digital financial services empower women to have greater control of their income and resources.** The Benazir Income Support Program, in Pakistan, is a large social cash transfer program that targets women below poverty level. Benefits are disbursed to recipients using smart cards, mobile phones, and debit cards. Thus, women receive their benefits conveniently, do not have to spend time going to and from a financial institution, and are able to leave some of the money received as savings in banks (Salman and Nowacka 2020). With digital payments and transfers, women have more privacy and control over their income flows, helping them overcome the challenges associated with their savings and investment decisions (Holloway, Niazi, and Rouse 2017). In Fiji, the government uses Help for Home, a mobile-enabled government-to-person social payments initiative, to expedite the distribution of social aid to rebuild homes of population affected by disaster (AFI 2021).

(iii) **Using digital financial services lowers the cost of payments, transfers, and remittances from family and friends in the event of disaster.** For example, Nationwide Microbank in Papua New Guinea, in partnership with Digicel, the largest telecommunications firm in the country, introduced MiCash, a mobile banking initiative.[19] MiCash allows clients to make financial transactions such as depositing and withdrawing money, sending and receiving money, and purchasing goods and services using a mobile phone (Erman, Obolensky, and Hallegate 2019). Orange money, the largest mobile banking company in Mali partnered with banks to allow bank customers to transfer money from their Orange Money electronic wallet to their bank account and vice versa. This facilitated access for remote clients (especially women) with no easy access to conventional bank counters (Haworth et al. 2016). In the Philippines, businesses, households, and individuals can transfer money across banks and e-money providers using the Instapay or Pesonet platforms. Aside from significantly reducing transaction costs in terms of time and money, digital financial services also help expand women's access to social networks that may provide support through remittances and transfers when a shock hits. A diversity of senders is facilitated as was observed in the Kenyan mobile money revolution (Lyons et al. 2020).

(iv) **Use of digital technology to assess credit risks may facilitate women's access to credit.** Some women are unable to access affordable credit because they cannot prove their creditworthiness and lack credit history and collateral. Alternative credit risk assessment mechanisms such as use of big data analytics tool, artificial intelligence-based technologies, or psychometrics to determine risks can facilitate women's access to credit (Salman and Nowacka 2020). By using data from mobile phone records, utility bills, and social media, credit scores may be generated. Such technology can facilitate women's access to credit and in the process build credit history. Following are some examples:

(a) LendMN, a financing company in Mongolia, developed a mobile application that uses artificial intelligence to assess creditworthiness, enabling the company to lend without asking for any collateral. Most of LendMN clients are women (Salman and Nowacka 2020).

(b) FINCA Bank in Georgia, with technical assistance from ADB, employed the digitization of field-based loan processing to provide credit access to rural MSMEs and smallholder farmers (mostly women). The bank engaged input and farm equipment suppliers as agents through which loans were disbursed and used to buy farm inputs and/or equipment and machineries. Borrowers' data are gathered and input to a credit decision tool that determines credit approval in a short time. Fast and convenient loan processing provided incentives for borrowers (mostly women) to avail of loans that were used to increase productivity and therefore enhanced their preparedness for disaster (Salman and Nowacka 2020).

(v) **Digital solutions also provide additional sources of income for women as entrepreneurs.** For example, in the Philippines, Grameen Foundation developed the Community Agent Network. It is a mostly female network (75% of its agents are women) that allows women who own or operate small convenience stores to become bank agents. Women are given point-of-sale devices and quick response codes to collect payments from clients. Similarly, the POSIBLE.Net point-of-sale device allows entrepreneurs in the Philippines to offer services such as utility bill payments, mobile loading, and ticketing. Women can act as agents and operate from their homes or businesses and earn a commission for every transaction (Salman and Nowacka 2020).

---

[19] The Pacific Private Sector Development Initiative, a regional technical assistance facility cofinanced by ADB, the Government of Australia, and the Government of New Zealand, has supported the expansion of MiCash.

Use of digital technologies enables financial institutions to be resilient in times of disasters. Digitization allows FSPs to continue servicing the financial needs of their clients during disaster events. A case in point is the experience of Cantilan Bank in the Philippines. Cantilan Bank was supported by ADB to migrate their core banking system to the cloud. When typhoon Odette hit several provinces on the eastern seaboard of the Philippines, most financial institutions, including branches of big commercial banks in the affected areas, took more than a month to restore services due to loss of connectivity and electricity. Cantilan Bank, however was able to restore services the day after the storm passed and provided the needed financial services to its clients. This underscores the importance of a digital ecosystem particularly during disasters.

As digital technologies reshape the financial landscape, women may be disadvantaged because of gender disparity in Internet and mobile phone access and usage. On average, women are 10% less likely to own a mobile phone than men—i.e., 84% of men compared to 74% of women own a mobile phone (World Bank 2018b). Further, even when women own a mobile phone, they use it less frequently than men (Holloway et al. 2017). In developing economies, 43% of men have access to both mobile phone and Internet while only 37% of women have access to both services. In some countries, such as Bangladesh, Ethiopia, and India, men are twice as likely as women to have access to both mobile phones and Internet services. If gender differences in access to mobile phone and Internet services are not appropriately addressed, use of digital financial services may lead to further gender inequality. Equal access to mobile phones and Internet services is however recorded in Colombia, the PRC, and South Africa (World Bank 2018b).

Women may also be disadvantaged with the emerging digitization due to inadequate knowledge of access, use, and security provided by digital solutions. Women need relevant and appropriate digital literacy to have a better understanding of the importance, relevance, and benefits of specific digital financial products and services including their specific use and application. With appropriate digital literacy training, women will be able to harness the efficiencies brought about by digital financial services and strengthen their economic resilience to disasters. The digital exclusion of women must be reduced, particularly for women in low-income and remote areas. To be able to benefit from digital financial services and participate fully in a digital economy, women should have access and capacity to use information and communication technology.

CONCLUSION AND RECOMMENDATIONS

Prevailing sociocultural norms and practices that lead to gender disparity in access to resources make women more vulnerable to disasters compared to men. Financial inclusion, through the availability and use of relevant financial instruments, can reduce these vulnerabilities and strengthen the economic resilience of women.

Most of the financial instruments available are designed to cater to both men and women. Very few are designed to cater to the unique and specific needs of women. Stakeholders therefore need to work together to facilitate and encourage the development, design, and promotion of relevant and appropriate financial products and services that will enhance women's resilience to disasters. This includes identifying distribution channels that are appropriate and familiar to women, particularly those in the low-income sector. This section lays out some recommendations for the specific sets of stakeholders.

Policy Makers and Regulators

Adopt gender-specific policies and strategies in national financial inclusion strategies. The dearth of financial instruments that are tailored to the specific and unique needs of women stems from the lack of a specific policy or strategy to deliberately enjoin FSPs to cater to the unique needs and preferences of women. The Global Microscope study reported that very few countries have a specific gender approach in their NFIS. Few countries have specific targets for closing the gender gap. Policy makers therefore need to adopt and implement strategic policy measures that target financial inclusion of women. Policies that support if not mandate the collection of sex-disaggregated data on the usage of financial services are important. Data to be collected will inform FSPs on women's preferences that could be used as a basis for designing and developing financial products and services that are relevant and tailored for women.

Establish policies that enable the use of bank accounts and digital financial services for distributing government assistance, particularly during disasters. Adopting and implementing such policies will result in more efficient and transparent distribution of financial assistance during disaster events. This will address women's mobility and time constraints. Policies that enhance women's access to national IDs should also be expedited. This will enable women to meet the simplified Know Your Customer requirements imposed by regulators and facilitate opening accounts with legitimate FSPs.

Adopt regulations that facilitate women's access to financial services. Regulations for formal financial institutions may also be tweaked and/or relaxed to enable FSPs to creatively cater to women's unique needs and preferences for financial products. Regulators may consider the following:

(i) Allow FSPs to offer no-frills savings accounts with relaxed Know Your Customer and low deposit requirements for opening accounts. This will enable and encourage women to open precautionary savings accounts.

(ii) Allow the use of agents that provide financial services particularly in remote areas. With the use of agents, women may not need to travel to access financial services. Agents who are familiar to women will also help encourage women to access and use financial services.

(iii) Institutionalize the provision of regulatory relief measures in times of disaster. These measures should be activated immediately and made available to FSPs during and even after a disaster.

(iv) Adopt an insurance regulatory framework that will encourage and enable private insurance providers to (a) participate in the provision of disaster risk insurance, and (b) design insurance products tailored to the unique needs of women in the event of shocks and disasters.

(v) Allow and encourage private insurance providers to partner with organizations that deal directly with women. These organizations can assist with and participate in the design, promotion, and distribution of insurance products relevant for women in times of disaster.

Development Partners

Support the incorporation of a strategic gender approach in country financial inclusion agendas. Development partners could support government efforts to identify and incorporate specific and relevant gender approaches to target the financial inclusion of women. Development partners may also assist governments to adopt specific strategies that encourage the collection of sex-disaggregated data on the different types of financial products and services offered by FSPs. The availability of sex-disaggregated data and information will help inform policy makers, regulators, and FSPs so they can formulate appropriate measures and design financial products and services relevant for women. The development community may also link with disaster risk reduction and management agencies to collect pertinent data and information on men and women during and after a disaster.

Support the establishment of financial mechanisms that will enable FSPs to meet the financing needs of women for enhanced economic resilience in times of disaster. To meet women's financing needs for recovery and rebuilding, some FSPs may need liquidity assistance. Such assistance is needed because, during disasters, the demand for loans usually increases while the collection and repayment of loans decrease. At such times, FSPs may need additional capital to support their operations. Development partners may also consider providing a portfolio guarantee for loans that women take for risk reduction or recovery. Portfolio guarantees could encourage FSPs to respond to women's financing needs in particular during disasters. Guarantees can absorb a portion of the default losses incurred by an FSP in return for a fee.

Conduct action research[20] **on the design of relevant and appropriate financial products for women's economic resilience to disasters.** Most financial products that strengthen economic resilience during disasters were designed without considering specific gender concerns and preferences. Action research will provide on-sight evidence, hands-on experience on the issues investigated, and knowledge that will help stakeholders to understand, determine, and identify the unique needs of women given the prevailing social and cultural norms. Grants to support this type of studies may focus on the appropriate design of financial products and services that will improve and strengthen women's economic resilience. Lessons learned from the studies may be used to design training and capacity building programs for FSPs.

Support the conduct of digital and financial literacy appropriate for and tailored to women. Many women may not be aware of or have the capacity to use the digital financial services even if they are made available and accessible. Thus, it is important to set up and implement financial and digital literacy programs for women. The programs should raise their awareness of financial concepts and products, inform them about the use of smartphones for digital financial services, and equip them to be able to do so.

Financial Service Providers

Incorporate the collection of sex-disaggregated data as part of FSPs' regular reporting systems. Collection of sex-disaggregated data will enable FSPs to better understand women clients' financial behavior patterns and preferences. This will provide an informed basis for developing and designing financial products for women. Such information will also help FSPs to appreciate women as a valuable market clientele. Sex-disaggregated data and information can be used to build an internal business case for developing programs for the women's market. With appropriate data and information, FSPs will be able to adopt a customer-centric approach that considers the different needs, preferences, behavior, and profitability of the women client segment.

Provide women access to secure and private savings accounts. Because women have a distinct preference for using savings rather than borrowing funds during emergencies, FSPs need to develop and customize financial products that meet women's needs and preference for precautionary savings. Savings products should be designed to assure women that their accounts are secure and confidential so they have control over the funds. With confidentiality and control, women are shielded from family pressure to share their savings with others. Building commitment into savings products (e.g., savings earmarked for specific purposes) also benefits women because it helps them meet specific expenditure goals.

---

[20] "Action research" refers to a wide variety of evaluative, investigative, and analytical research methods designed to diagnose problems to develop practical solutions to address them quickly and efficiently. Action research may be applied to situations that researchers want to learn more about and improve. The general goal is to create a simple, practical, repeatable process of iterative learning, evaluation, and improvement that leads to increasingly better results. (Based on definition provided in the Glossary of Education Reform, https://www.edglossary.org/action-research).

Collaborate with organizations that work directly with women as clients and/or as members. When designing and developing financial products and services, FSPs may tap women's organizations for advice. They may also be used as distribution and delivery channels of financial services. Women-friendly distribution and delivery channels are particularly relevant and important in providing insurance products for women in the informal sector. The collective purchase of insurance (especially for disaster relief and income protection) by organizations and/or associations that are working closely with women should be considered.

Adopt innovative digital technologies to provide access to unserved women and employ alternative forms of credit assessment. Provision of digital technology can reduce the transaction cost, time, and mobility constraints women face in accessing financial services. Lenders may opt to use alternative credit assessment methods that use big data collected from various digital platforms (e.g., social media) to determine women's creditworthiness. This may include data or information collected and collated through mobile phone use (e.g., timeliness of bill payments, extent of contact list, social circle, range of activities engaged in). This information can be used to determine women's creditworthiness without requiring collateral and credit histories.

Include women staff in designing and developing financial products. This will help to ensure that the products meet women's needs. Employing women as account officers and/or agents who will directly deal with women clients will also facilitate access and encourage women to use financial products and services.

The use of appropriate financial products and services can help prepare women for disaster events and provide them safety nets that can bolster their disaster coping capacities. By working together, key stakeholders in the sector can promote the design and use of relevant and appropriate financial instruments. The instruments should take into consideration the differential impact of disasters on men and women and thereby enhance the economic resilience of women to disasters.

REFERENCES

Alliance for Financial Inclusion (AFI). 2009. *First Annual Global Policy Forum. A Marketplace of Ideas.* https://www.afi-global.org/wp-content/uploads/publications/2009report.pdf.

———. 2021. *Disaster Resilience through Financial Inclusion: The Role of Financial Regulators in Disaster Risk Reduction.* https://www.afi-global.org/wp-content/uploads/2021/02/AFI_IGF_disaster-resilience_AW_1304.21_digital.pdf.

Arnold, J. 2021. The Case for a Gender-Intelligent Approach: An Opportunity for Inclusive Fintechs. Center for Financial Inclusion. 28 January. https://www.centerforfinancialinclusion.org/the-case-for-a-gender-intelligent-approach-an-opportunity-for-inclusive-fintechs.

Ashraf, N., D. Karlan, and W. Yin. 2006. *Female Empowerment: Impact of Commitment Savings Product in the Philippines.* Economic Growth Center Yale University. Center Discussion Paper No. 949. Yale University. http://www.econ.yale.edu/growth_pdf/cdp949.pdf.

Asian Development Bank. 2020. *Enhancing Women-Focused Investments in Climate Disaster and Resilience.* Manila. https://dx.doi.org/10.22617/TCS200140-2.

Bangko Sentral ng Pilipinas (Central Bank of the Philippines). n.d. 213 Basic Deposit Account. https://morb.bsp.gov.ph/213-basic-deposit-account/.

———. n.d. Inclusive Finance—Financial Inclusion. https://www.bsp.gov.ph/Pages/InclusiveFinance/InitiativesAndRegulation.aspx.

Banqin. 2020. A Look into Financial Inclusion. https://www.getbanqin.com/post/a-look-into-financial-inclusion-in-thailand.

Brown, W., and G. Nagarajan. 2000. *Bangladeshi Experience in Adapting Financial Services to Cope with Floods: Implications for the Microfinance Industry.* Development Alternatives Inc. http://www.gdrc.org/icm/disasters/bangladeshi_experience_in_adapting_financial_services.pdf.

Business World. 2021. Basic Deposit Accounts Climb to 7M as of March. 9 November. https://www.bworldonline.com/basic-deposit-accounts-climb-to-7m-as-of-march/.

Buvic, M., and M. O'Donnell. 2016. Gender and Financial Inclusion: With a Nudge and a Twist. Center for Global Development. Blog. 16 April. https://www.cgdev.org/blog/gender-and-financial-inclusion-nudge-and-twist.

Center for Global Development. n.d. *Mindful Saving: Exploring the Power of Savings for Women* (Factsheet). https://www.cgdev.org/sites/default/files/mindful-saving-exploring-power-savings-women-factsheet.pdf.

Concern Worldwide. 2015. What is Disaster Risk Reduction (and why does it matter). Ireland. 12 November. https://www.concern.net/news/what-is-disaster-risk-reduction-and-why-does-it-matter.

de Vera, B. 2021. Number of Filipinos Covered by Microinsurance Drops to 46.97 Million in Q1. *Inquirer.net.* 29 October. https://business.inquirer.net/333478/number-of-filipinos-covered-by-microinsurance-drops-to-46-97-million-in-q1.

Dhakal, N., N. Simkhada, and M. Ozaki. 2019. Microfinance for Disaster Recovery: Lessons from the 2015 Nepal Earthquake. *South Asia Working Paper Series.* No. 65. Manila: Asian Development Bank. https://www.adb.org/sites/default/files/publication/502256/swp-065-microfinance-disaster-recovery-nepal-earthquake.pdf.

Economist Intelligence Unit (EIU). 2019. *Global Microscope 2019: The Enabling Environment for Financial Inclusion.* New York, NY.

————. 2020. *Global Microscope 2020: The Role of Financial Inclusion in the COVID-19 Response.* New York, NY.

El Zoghbi, M. 2020. Advancing Women's Financial Inclusion: What Does the Global Microscope Tell Us? *Center for Financial Inclusion Blog.* 6 January. https://www.centerforfinancialinclusion.org/advancing-womens-financial-inclusion-what-does-the-global-microscope-tell-us.

Erman, A., M. Obolensky, and S. Hallegate. 2019. *Wading Out the Storm.* Washington, DC: Global Facility for Disaster Reduction and Recovery, World Bank Group. https://documents1.worldbank.org/curated/en/626361565186647096/pdf/Wading-Out-the-Storm-The-Role-of-Poverty-in-Exposure-Vulnerability-and-Resilience-to-Floods-in-Dar-Es-Salaam.pdf.

Erman, A., S. Robbe, S. Thies, K. Kabir, and M. Maruo. 2021. *Gender Dimension of Disaster Risk and Resilience.* Washington, DC: Global Facility for Disaster Reduction and Recovery, World Bank Group. https://openknowledge.worldbank.org/bitstream/handle/10986/35202/Gender-Dimensions-of-Disaster-Risk-and-Resilience-Existing-Evidence.pdf?sequence=1&isAllowed=y.

German International Cooperation. 2020. *The Landscape of Climate and Disaster Risk Insurance in South and Southeast Asia and Oceania.* https://mefin.org/docs/GIZ-Climate-and-Disaster-Risk-Insurance-CDRII-Landscape-Study.pdf.

Global Partnership for Financial Inclusion. G20 Financial Inclusion Indicators. https://datatopics.worldbank.org/g20fidata/.

Government of the Philippines. 2010. *Philippine Disaster Risk Reduction and Management Act of 2010,* Republic Act No. 10121. https://www.officialgazette.gov.ph/2010/05/27/republic-act-no-10121/.

Hallegate, S., A. Vogt- Schilb, M. Bangaloor, and J. Rozenberg. 2017. *Unbreakable: Building Resilience of the Poor in the Face of Natural Disaster.* Climate Change and Development Series. Washington, DC: World Bank Group. https://openknowledge.worldbank.org/bitstream/handle/10986/25335/211003ovEN.pdf.

Haworth, A., C. Frandon-Martinez, V. Fayolle, and C. Simonet. 2016. *Climate Resilience and Financial Services.* Building Resilience and Adaptation to Climate Extremes and Disasters Knowledge Manager. https://www.preventionweb.net/publications/view/47933.

Holloway, K., Z. Niazi, and R. Rouse. 2017. *Women's Economic Empowerment Through Financial Inclusion.* Innovations for Poverty Action. https://www.poverty-action.org/sites/default/files/publications/Womens-Economic-Empowerment-Through-Financial-Inclusion.pdf.

International Center for Climate Change and Development. 2014. *Proceedings of the GIBIKA Workshop and Seminar on Index-Based Insurance.* https://icccad.net/wp-content/uploads/2015/12/ICCCAD-GIBIKA-IBI-Second-Workshop-Report.pdf.

Intergovernmental Panel on Climate Change (IPCC). 2012. *Managing the Risks of Extreme Events and Disasters to Advance Climate Change Adaptation. Special Report of the IPCC.* Geneva. https://www.ipcc.ch/report/managing-the-risks-of-extreme-events-and-disasters-to-advance-climate-change-adaptation/.

Iyer, R. 2019. Financial Inclusion in India is Soaring: Here's What Must Happen Next. World Economic Forum. 14 January. https://www.weforum.org/agenda/2019/01/financial-inclusion-in-india-is-soaring-heres-what-must-happen-next/.

Jacobsen, K., A. Marshak, and A. Griffith. 2009. *Increasing Resilience of Disaster Affected Population.* Medford, MA: Feinstein International Center.

Karlan, D. et al. 2011. *Evaluating the Savings for Change Program in Mali.* Innovation for Poverty Action. Washington, DC: Global Facility for Disaster Reduction and Recovery, World Bank Group.

Lyons, A. C., J. Kass-Hanna, F. Liu, A. Greenlee, and L. Zeng. 2020. Building Financial Resilience through Financial and Digital Literacy. *Asian Development Bank Institute Working Paper Series.* No. 1098. Tokyo: Asian Development Bank Institute. https://www.adb.org/sites/default/files/publication/574821/adbi-wp1098.pdf.

Martinez, J. 2017. Financial Inclusion in Malaysia: Distilling Lessons from Other Countries. *The Malaysia Development Experience Series.* Washington, DC: World Bank. https://documents1.worldbank.org/curated/en/703901495196244578/pdf/115155-WP-PUBLIC-GFM08-68p-FIpaperwebversion.pdf.

Miles, K., and M. Wiedmaier-Pfister, M. 2018. *Applying a Gender Lens to Climate Risk Finance and Insurance.* InsuResilience Global Partnership. https://www.insuresilience.org/wp-content/uploads/2018/11/insuresilience_applygender_181128_web.pdf.

Miles, K., M. Wiedmaier-Pfister, and C. Dankmeyer. 2017. *Mainstreaming Gender and Targetting Women in Inclusive Insurance: A Compendium of Technical Notes and Case Studies.* German International Cooperation. https://www.ifc.org/wps/wcm/connect/4dbd983e-2ecd-4cde-b63e-191ffb2d48e6/Full+Women+%26+Inclusive+Insurance+BMZ_Web.pdf?MOD=AJPERES&CVID=lK1xhtq.

Moore, D., Z. Niazi, R. Rouse, and B. Kramer. 2019. *Building Resilience through Financial Inclusion: A Review of Existing Evidence and Knowledge Gap.* Innovation for Poverty Action. https://www.poverty-action.org/sites/default/files/publications/Building-Resilience-Through-Financial-Inclusion-January-2019.pdf.

Pantojan, E. 2002. *Microfinance and Disaster Risk Management: Experiences and Lessons Learned.* Prevention Consortium. Washington, DC: World Bank. https://www.gdrc.org/icm/disasters/microfinance_drm.pdf.

Patel, T., and R. Nanavaty. n.d. Case Study 4: The Experience of SEWA. *Invest to Prevent Disaster.* https://www.unisdr.org/2005/campaign/docs/case-study-4-Microfinance-and-Disaster-Mitigation-sewa.pdf (accessed 23 July 2021).

Plata, G. 2021. Global Microscope on Financial Inclusion: How does Your Country Rank? Washington, DC: Inter-American Development Bank. https://www.iadb.org/en/improvinglives/global-microscope-financial-inclusion-how-does-your-country-rank.

Salman, A., and K. Nowacka. 2020. Innovative Financial Products and Services for Women in Asia and the Pacific. *Sustainable Development Working Paper Series.* No. 67. Manila: Asian Development Bank. https://www.adb.org/sites/default/files/publication/576086/sdwp-67-financial-products-services-women-asia-pacific.pdf.

Southeast Asia Disaster Risk Insurance Facility. www.seadrif.org.

Swiderek, D., and J. Wipf. 2015. *Aiding the Disaster Recovery Process: The Effectiveness of Microinsurance Service Providers' Response to Typhoon Haiyan.* Microinsurance Network. http://www.microinsurancecentre.org/resources/documents/policyholder-value-of-microinsurance/aiding-the-disaster-recovery-process-the-effectiveness-of-microinsurance-service-providers-response-to-typhoon-haiyan.html.

SWIFT. Know Your Customer. https://www.swift.com/your-needs/financial-crime-cyber-security/know-your-customer-kyc/meaning-kyc#:~:text=illicit%20criminal%20activities.-,Know%20Your%20Customer%20(KYC)%20standards%20are%20designed%20to%20protect%20financial,of%20funds%20is%20legitimate%3B%20and.

Tighe, K. n.d. The Role of Microfinance and Microinsurance in Disaster Management. *Advanced Center for Enabling Disaster Risk Reduction.* Research Brief. No. 2. https://www.preventionweb.net/files/20073_researchbrief2roleofmicrofinanceind.pdf.

Trivelli, C. et al. 2018. *Financial Inclusion for Women: A Way Forward.* Consejo Argentino Para Las Relaciones Internacionales (CARI). https://www.g20-insights.org/wp-content/uploads/2019/10/Financial-Inclusion-for-Women-Final.pdf.

Ullah, I., and M. Khan. 2017. Microfinance as a Tool for Developing Resilience in Vulnerable Communities. *Journal of Enterprising Communities: People and Places in the Global Economy.* 11(2). pp. 237–57. doi: 10.1108/JEC-06-2015-0033.

Women's World Banking. 2018. How to Create Financial Products that Win with Women. 4 September 2018. https://www.womensworldbanking.org/insights-and-impact/how-to-create-financial-products-that-win-with-women/ (accessed 22 March 2021).

World Bank. 2018a. Global Findex. Data. Washington, DC. https://globalfindex.worldbank.org/ (accessed 12 February 2021).

————. 2018b. Global Findex. The Global Findex Database 2017. Washington, DC. https://globalfindex.worldbank.org/ (accessed 12 February 2021).

————. 2018c. Overview. *Global Findex Report.* Washington, DC. https://globalfindex.worldbank.org/basic-page-overview.

Lightning Source UK Ltd.
Milton Keynes UK
UKHW050913160922
408923UK00005B/237